WELDABILITY OF STEEL

Prepared by K. G. Richards, B.Sc., A.C.G.I.

THE WELDING INSTITUTE

Abington Hall, Abington, Cambridge CB1 6AL

Published November 1967
Reprinted January 1970
Second reprint January 1972
Third reprint November 1976
Fourth reprint July 1980

CONTENTS

3	**Foreword**
4	**Introduction**
5	**Factors influencing weldability**
5	Heat affected zone
5	Material composition
7	Residual stress
7	Welding procedure
8	**Potential welding problems**
9	**Weld metal hot cracking**
9	Restraint
10	Weld shape
11	Composition
12	**Heat affected zone burning and hot tearing**
13	**Lamellar tearing**
14	**Heat affected zone hydrogen induced cracking**
15	Thickness and type of joint
16	Composition
16	Type of electrode or welding process
17	Energy input and preheating temperature
17	Steel classification
19	**Heat treatment cracking**
20	**Corrosion**
20	Mechanism of corrosion
20	Crevice corrosion
21	Weld decay
21	Stress corrosion
22	Corrosion fatigue
23	**Factors affecting joint efficiency**
23	Static strength
23	Brittle fracture
25	Fatigue strength

FOREWORD

Every year the Institute's staff are concerned in a number of investigations of failures in welded constructions of various kinds. Most of these could have been avoided if greater attention had been paid to some of the special problems arising from the use of welding as a joining process and if existing knowledge had been properly applied. Even if the expert knowledge is not always available in every company concerned, advice can always be obtained from the Institute's staff and indeed from other sources, but the real difficulty seems to be in a fairly widespread lack of awareness in the first place that there may be a problem on which it would be a wise precaution to seek expert advice.

Most welding processes in industrial use require the application of heat. With few exceptions – such as friction welding – melting and resolidification occurs in most welding processes. The parent metal adjacent to the weld even if it has not been melted will experience heating to a high temperature followed by cooling. As a consequence, a number of complex chemical and metallurgical changes may occur which may produce quite profound changes in the mechanical properties and behaviour of the parent metal and the weld originally produced in manufacture by careful control of chemical composition, mechanical working and heat treatment. Sometimes these changes may be so drastic that a sound welded joint cannot be produced with the welding process envisaged. Even a small initial crack that escapes non-destructive examination may cause complete failure of the structure if it is propagated as a brittle fracture or as the result of fatigue under repeated loading.

It must be obvious that there is little point in choosing a material for a particular duty on account of its strength, hardness, toughness, wear-resistance or other valuable property if these properties are greatly diminished or even destroyed as a result of welding. It has not been unknown for failures to occur as the result of using a special and perhaps rather expensive material whose valuable properties are greatly impaired by welding when a cheaper and more readily weldable material would have given quite satisfactory service.

The first design memorandum issued by the Institute a few months ago dealt with joint preparations. This second one which will be followed by others, deals with some of the problems of a metallurgical character which may arise in welding and which the designer should bear in mind when specifying materials. It is intended for use by engineers, particularly those in design and drawing offices, and it is hoped that its contents are presented in such simple form that it will be readily understandable even to those whose knowledge of metallurgy is a little rusty.

It is not the purpose of this memorandum to tell designers what materials they should use or should avoid, or to give detailed guidance on the weldability of materials made to a multitude of different specifications. This would be an impossible task. Its real purpose is to create a higher degree of awareness of the problems that might arise from the choice of materials for welding and the factors that have to be considered in making the choice. It will, it is hoped, assist designers to make a better choice than is sometimes the case and to encourage designers to seek expert advice on the precautions to be taken where, for specific reasons, there is no choice and a material may have to be used that could lead to trouble or even fracture in service.

Any comments, suggestions and proposals for improvement of the information presented in this memorandum will be most gratefully received for incorporation in future editions.

R. WECK,
Director General

INTRODUCTION

Unfortunately neither 'weldability' nor 'steel' can be explained in a simple precise definition. This handbook describes in general terms the weldability of steel from the designer's viewpoint. Weldability is not an absolute term, but is relative. It takes on different meanings depending on whether one is a designer, a welding engineer or a metallurgist. To add to the problem 'steel' covers a host of alloys with different combinations of chemical elements.

What does the average designer understand by 'weldability'? He would probably say: "The ease with which a material may be joined, by as many as possible of the common welding processes, to produce a joint having properties suitable for the conditions of service to be imposed upon it." The desired properties include static strength, fatigue strength, notch ductility, corrosion resistance and appearance, any of which may be required at low or elevated temperatures. To establish the necessary degree of weldability it is imperative that the designer should know what conditions of service he is trying to meet. For example, it is a waste of effort to express the desired weldability in terms of static strength when the only requirement is the appearance or colour matching of the weld metal.

Steels range from plain carbon mild steel, through carbon-manganese and low alloy steels, to high strength or corrosion resistant alloy steels. Each type or group of steels is covered by a specification which is normally confined to chemical composition and mechanical properties. When a designer chooses a particular specification for his steel he expects that any material supplied under that specification will be equally weldable. Unfortunately, for reasons which will be explained later, this is not necessarily true, and in fact steel may be supplied within the normal limits of a specification which may range from readily weldable to unweldable without special precautions.

It should be apparent, that the designer must try to select the material which meets the properties he requires when welded, but with the minimum of problems during welding. He must make himself aware of what these problems may be and how they are affected by his design.

FACTORS INFLUENCING WELDABILITY

HEAT AFFECTED ZONE

Before embarking on a description of particular difficulties which can arise during welding, it may be useful to consider briefly what happens to a material when it is welded and what are the main factors controlling its weldability.

Fusion welding processes, which are those most commonly used for joining steel, require the local application of heat energy in order to bring the material to a temperature at which it will fuse. For steels this is about 1400 – 1500°C. This energy is dissipated into the surrounding atmosphere and into the parent plate material either side of the weld.

Consider a butt weld joining two pieces of steel plate made by the manual metal arc process, in which the heat is supplied by means of an arc struck between a coated steel electrode and the plates. The electrode itself melts to provide weld metal to fill the gap between the two prepared edges of the plates. During welding a temperature gradient is set up in the material ranging from the fusion temperature in the weld pool down to the initial plate temperature some distance away from the weld. The material adjacent to the weld, even though it does not melt, undergoes a severe thermal cycle as a result of welding and because of this, microstructural changes may occur in those regions or Heat Affected Zones (HAZ). The extent of these changes in microstructure will depend mainly on the composition of the material and the rate of heating and cooling during the thermal cycle. The rate of cooling in the Heat Affected Zones will depend on the size of the job, particularly the thickness of the material being joined, and on the rate of heating produced by the welding process employed. For a constant heat input, the rate of cooling will increase rapidly with increasing plate thickness. Cooling rates well in excess of those produced by water quenching can be achieved. If the material is of a composition which is hardenable, then a hardened HAZ will be produced by the cooling rates normally associated with welding.

This hardening of the HAZ microstructure is just one of the potential problems which may arise when steel is welded. These will be described in more detail later. This point serves to illustrate that

Prepared plates

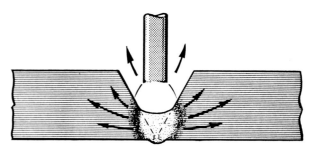
Heat flow during welding

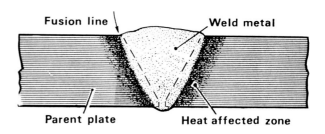
Parent plate — Heat affected zone — Fusion line — Weld metal

the composition of the material is one of the main variables affecting its weldability.

The original properties of the material produced by careful alloying and heat treatment may be drastically altered by the application of the welding heat.

COMPOSITION

The specification for a structural steel normally controls the limits of composition only for the main elements carbon, manganese, silicon, sulphur and phosphorus. For a low alloy steel, nickel, chromium and molybdenum are also specified. The steel is normally produced by casting the molten steel from a ladle into an ingot, then rolling the ingot into plate or other form. The mechanical properties of the plate depend to some extent both on the rolling temperature and on the cooling rate from the

Section of fully killed steel ingot showing segregation pattern

rolling temperature. The thicker the section produced, the higher the rolling temperature and the slower the cooling rate after rolling or normalising,* therefore the lower the mechanical properties for a given composition. Mechanical working such as rolling or forging of the material improves the mechanical properties to some degree but since a specification covers the production of material over a range of plate thickness this means that the range of chemical composition has to be wide enough to achieve the minimum mechanical properties for each thickness. In other words, because of the smaller amount of rolling and slower cooling rate from the rolling temperature, to get the same mechanical properties of yield point, ultimate strength and percentage elongation in thick plate, the quantity of strengthening elements, such as carbon and manganese, has to be higher than in an equivalent thin plate.

Chemical analysis during manufacture is usually carried out while the liquid steel is in the ladle. During solidification of the ingot, segregation of some of the elements occurs because the material solidifies progressively over a range of temperature and not at one particular temperature. Thus the composition at the centre of the ingot may be different from that at the outside. This segregation is reduced by hot working and rolling of the ingot but even on the finished plate the composition can vary considerably across the width and thickness of the plate. The effect is more apparent on thick plate, firstly, because the thicker the section the larger the ingot required to produce it, which means a slower cooling rate giving more segregation, and secondly, because there is less rolling required to produce a thick plate and so the segregation is not so much reduced.

Another factor affecting the composition of a steel is the presence of small amounts of impurities or trace elements. Most of these are not normally covered by a specification, but it is becoming increasingly apparent that small quantities of tin, copper, arsenic and aluminium, for example, can have a considerable influence on the weldability. The effect is intensified on thicker sections because of concentration of these elements by segregation.

The composition of the weld pool is another variable which is difficult to predict as it is made up partly from the filler metal used and partly from the parent metal. The proportion of parent metal in the melted pool, or dilution, depends on the type of joint and the characteristics of the welding process, but

---

* Normalising is a heat treatment intended to improve certain mechanical properties of a steel by producing a refinement of the grain structure. It involves heating the steel to a high temperature (the precise value will depend on the particular steel composition) followed by cooling in still air.

the problems which may arise and the mechanical properties of the weld metal will depend, at least in part, on the composition of the parent metal.

It is clear that difficulties arising during welding will become more severe as the thickness of material is increased. This is true not only of the metallurgical factors but also of other contributory factors such as welding procedure and residual stress.

RESIDUAL STRESS

A residual stress system is a system of internal stresses in a body which can exist in the absence of any external loading and which is balanced within the body itself. If there is an area of tensile residual stress in any cross section at one part of the body, there must be a residual compression at some other part. Residual stresses arise as a result of permanent plastic strain in the material and the magnitude of the stress at any point depends on the amount of strain at that point. In a welded structure these strains are the result of the local heating and cooling cycles associated with welding, and especially the shrinkage of the weld metal. As the weld metal cools from the fusion temperature it tries to shrink, but is restrained by the parent material surrounding it. This shrinkage has to be largely accommodated by plastic straining of the weld metal. As a result of this plastic strain it follows that the level of residual stress in the weld metal must be at the yield point of the material unless it is modified by any subsequent treatment such as heating or stretching.

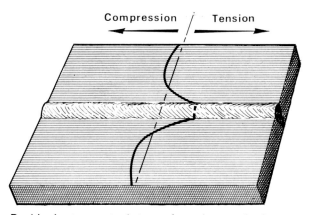

Residual stress at plate surface due to single weld bead

The amount of strain which has to be accommodated by the weld metal will depend on the degree of restraint applied by the parent plate. This restraint will be greater as the size and thickness of the plates being joined is increased. The plastic deformations which give rise to residual stress may also produce metallurgical effects which alter the physical properties of the material. When certain steels are deformed plastically, particularly under the action of heat and time, the ductility and notch toughness may be considerably reduced and the steels are said to be 'strain aged'. This effect may also be accompanied by 'precipitation', which means that some of the constituents of the material tend to separate out of solution and possibly produce embrittlement.

Heat treatment subsequent to welding may be used to alleviate some of these effects, but, with certain steel compositions, more harm may be done to the material especially if the treatment is not properly applied. Precipitation embrittlement may result from a stress relieving heat treatment and give rise to cracking in the weld metal or heat affected zone.

WELDING PROCEDURE

The most practical factor having considerable influence on the weldability of steel is the welding procedure and the welding process employed in that procedure. The fusion welding processes normally used for welding steel are:—

Tungsten inert gas (TIG) process
Manual metal arc process with coated electrodes
Gas shielded consumable metal arc processes
Continuous covered electrode processes
Submerged arc process
Electroslag processes

All these processes, with the exception of electroslag welding, produce heat by means of an electric arc established between an electrode and the material being joined. The electroslag process uses electrical resistance heating of a slag pool. The main difference between the processes is in the concentration of heat produced by them and the above list is approximately in order of increasing heat input. Obviously each process can cover a wide range of heat input so that a considerable degree of overlapping is involved in this respect.

For a given plate thickness, the greater the energy input per unit length of weld, the slower the subsequent cooling rate will be in the weld metal and heat affected zone. Since weldability can be very much a function of cooling rate then the choice of welding process is important. The choice of process is affected to a great extent by practical and economic considerations and more often than not situations arise where a process has to be used which will not result in an adequate cooling rate. In these circumstances it is necessary to adopt a welding procedure which will minimise the deficiencies of the welding process itself. This may involve the application of preheat, interpass and post heat to the joint being made. In many cases preheat alone may be sufficient.

As the names suggest these techniques involve the application of heat to the parts to be joined, before, during and subsequent to welding, in addition to the heat supplied by the welding process itself. The method of applying this heat will depend on the size and thickness of the job, the convenience and cost of the methods available and the degree of preheat etc. required. The most usual methods are furnace heating, direct flame heating using gas torches, and electrical resistance heating.

Whatever method of heating is employed it is essential that the heat is applied for a sufficient time to ensure that the joint to be made is brought to the required temperature throughout its thickness and for some distance away from the site of the joint. This applies particularly when preheat is being carried out by means of gas torches. Where temperature indicating crayons are used these should be either applied to the opposite side of the plate from the torch or the torch should be removed for a minute or so to allow the temperature to stabilise before checking. Since techniques of this kind are costly, there is no point in wasting money by applying them improperly.

The influence of preheat etc. on the weldability of steel will be described in more detail later, but one effect is the reduction of the hydrogen content of the weld metal and HAZ. The hydrogen concentration in the weld is determined principally by the type of electrode chosen or the type of process used. The coating of covered electrodes for example, consists of a mixture of mineral particles bonded together and usually containing free moisture and combined water of crystallisation. In the heat of the welding arc water molecules decompose to give hydrogen, which then dissolves in the weld metal. The hydrogen dissolved in this way is relatively mobile and can diffuse from the weld metal to the heat affected zone both during cooling and afterwards at room temperature. Different types of coating and different types of process will produce different amounts of hydrogen in the weld metal. This may range from 20 to 30 ml/100g for weld metal deposited with rutile coated electrodes down to 1 to 2 ml/100g using the gas shielded processes. Deposits made with 'low hydrogen' electrodes or the submerged arc process will usually be somewhere in between at about 10 ml/100g.

These figures presume that electrodes and fluxes are properly dried and stored before use and that the prepared joint is clean and dry. Significant amounts of hydrogen can be introduced into a weld if the joint is dirty, rusty or contaminated with oil or if filler wire is used which has a poor surface finish and has become contaminated with drawing lubricants.

Thus any degree of control over the weldability of a material ultimately depends on continuous attention to welding procedure.

POTENTIAL WELDING PROBLEMS

Problems which can arise during welding of steels may be divided into two categories:

1 Problems of weld metal or heat affected zone cracking directly attributable to the welding process.

The first category covers:
  weld metal hot cracking
  heat affected zone burning and hot tearing
  lamellar tearing
  heat affected zone hydrogen induced cracking

2 Problems which may occur at welded joints during service.

The second category:
  heat treatment cracking
  corrosion
  factors affecting joint efficiency

Not all the possible problems can be included under these headings, but those covered are the most likely to be found in practice.

WELD METAL HOT CRACKING

Longitudinal hot crack in manual butt weld

Weld metal hot cracking or solidification cracking to be more precise, is one of the more common types of defect found in welded structures. The term 'hot cracking' is derived from the fact that the fracture surfaces of the cracks are characterised by a blue appearance resulting from their formation and subsequent oxidation at high temperature (above 1200°C). Weld metal cracks may generally be classified as longitudinal, transverse, crater and hairline cracks which are usually associated with slag inclusions. The most usual type is the longitudinal crack through the 'throat' or centreline of a weld and which may occur in isolation or as an extension of a crater crack. These cracks may run to several feet in length in severe instances.

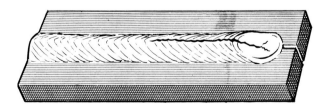

The main factors which control the susceptibility to hot cracking are:
 restraint
 weld shape
 material composition

Cracking may result from any one of these factors but will be more likely if two or more effects are operating at the same time.

RESTRAINT

As mentioned previously the shrinkage of weld metal from its fusion temperature produces residual stresses the magnitudes of which depend to a large extent on the degree of restraint afforded by the surrounding material. Under conditions of high restraint the hot strength and ductility of the cooling weld metal may become exhausted resulting in the formation of a crack. Restraint cracking, as it is sometimes known, is usually associated with small weld runs on joints in thick material as, under these conditions, all of the shrinkage has to be accommodated by plastic deformation of the weld metal itself.

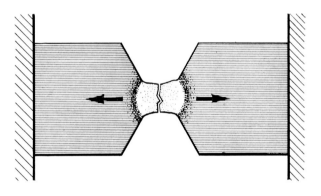

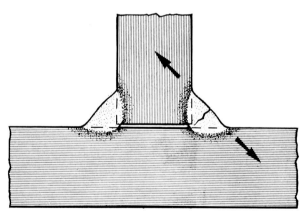

The fact that the cracks usually occur through the centreline of the weld is due to the solidification pattern of the weld metal. Since the parent material is much cooler the weld solidifies from the fusion lines inward which means that the centre of the weld is the last to solidify and is thus that much weaker at the temperature at which cracking occurs.

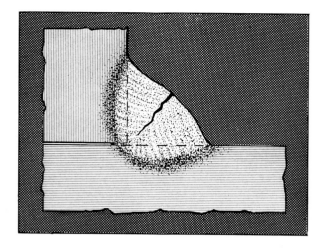

Occasionally, transverse hot cracks may occur where conditions give high restraint in the longitudinal direction of the weld. These cracks may propagate into the parent plate.

To minimise the effects of restraint it is necessary to give careful consideration to the joint conditions and the degree of restraint opposing movement of the parts caused by weld shrinkage. The order of assembly should be such as to allow shrinkage to take place as freely as possible consistent with any control of distortion required. The size of individual weld runs should be large enough to withstand shrinkage stresses particularly when welding thick sections. It may be desirable in certain circumstances to use electrodes of higher ductility, which are thus more resistant to cracking. Basic coated class 6 electrodes (low hydrogen type) are usually effective in this respect but in severe instances it may be necessary to use an austenitic type electrode.

WELD SHAPE

The effect of weld shape on the tendency to hot crack formation can largely be explained in terms of the reduction of area at critical sections of the weld metal. Very concave fillet welds or lack of root penetration both give sections which are less resistant to cracking especially under conditions of restraint. Similar effects may occur in the root runs of butt welds and also due to bad fit up between the parts being joined. Proper attention to edge preparation and correct fit up are necessary and also control of welding conditions and technique to ensure adequate penetration and correct weld shape.

Hot crack in fillet weld associated with lack of penetration

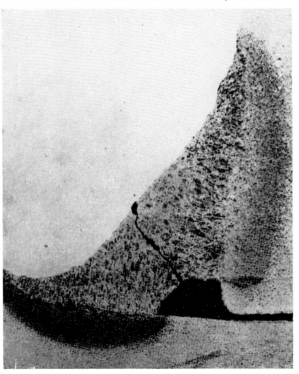

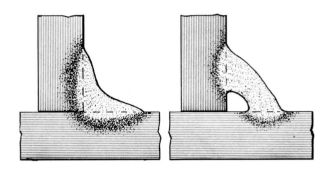

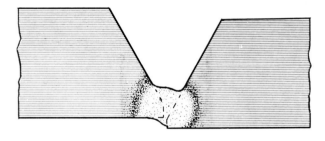

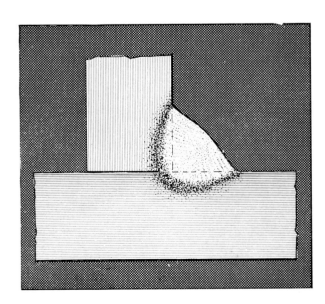

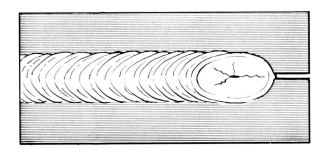

Another example largely attributable to weld shape is the tendency for cracking at the crater formed when the welding arc is broken off at the end of a run. If a deep crater is allowed to form, small cracks may occur radiating from the centre and these may extend into longitudinal cracks either in the same run or subsequent run of weld metal. Deep craters may be avoided by suitable welding technique. The arc should not be broken abruptly at the end of the run but the electrode should be traversed back over the deposit for a half to one inch before breaking the arc. Where high current mechanised welding is used, deep craters are particularly likely and therefore run on and run off plates should be used where possible, so that the machine starts and finishes its run clear of the joint.

Centre-line crack in root pass in 40°V preparation in 3 in. thick plate using high current CO_2 process

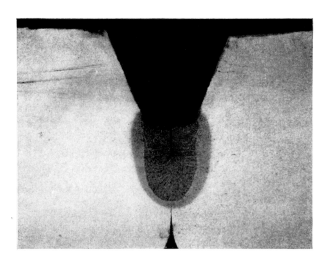

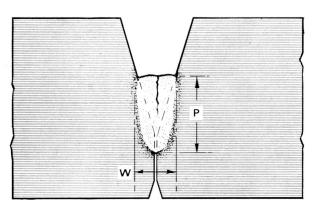

Some modern automatic machines employ a crater filling device in the electrical circuit.

A combination of restraint and an unfavourable solidification pattern may produce centreline cracking when using the deep penetration processes (*e.g.* high current CO_2 process). This is usually brought about by the use of too narrow a joint preparation which produces a weld bead having a high ratio of penetration to bead width. A ratio of P/W of less than 1·2 is usually sufficient to produce a crack free weld and deep, narrow preparations should be avoided.

MATERIAL COMPOSITION

Longitudinal centre-line crack in submerged arc root run in 2 in. thick plate caused by sulphur segregation

Weld metal contains certain elements which, if present in sufficient quantity, may segregate to form low melting point films between the solidifying crystals of metal. Under the action of solidification stresses these films may open up into cracks. The most common source of this problem is the presence of sulphur, the effect of which is further aggravated by increasing carbon content. Most of the elements and impurities giving rise to difficulty are contained in the parent plate and thus the amount of these elements present in the weld metal depends on the degree of dilution or mixing of the parent and electrode material. This, in turn, is dependent on the joint preparation and welding process used.

Sulphur forms films of iron sulphide which may be avoided by the presence of sufficent manganese to form small globules of manganese sulphide. Free cutting steels containing 0·25% S can be welded satisfactorily provided that the electrodes used contain sufficient manganese to absorb the sulphur. Basic coated class 6 electrodes are normally effective for this application.

The effectiveness of manganese in absorbing sulphur into spherical inclusions may be almost nullified if significant amounts of aluminium are present in the electrode wire or coating. This is particularly noticeable when using the CO_2 process, where wires containing silicon and manganese only as deoxidisers may be used successfully on free cutting steels whereas wires containing silicon, manganese and aluminium give rise to low melting point films and hence hot cracking.

Most structural steels nominally contain less than 0·06% S but this figure may be increased significantly by segregation effects along with carbon and other elements. This may increase the risk of hot cracking in the root runs of double V preparations, for example. The effect of this may be accentuated even more on very thick sections where the normal composition limits are allowed to rise.

HEAT AFFECTED ZONE BURNING AND HOT TEARING

Burning and hot tearing in heat affected zone of an oxy-acetylene weld in mild steel

Burning and hot tearing in the heat affected zone of welded joints is not such an easily recognisable phenomenon as hot cracking of the weld metal but it is described here because, to the layman at least, it is closely related to sulphur induced hot cracking.

When steel is heated close to its fusion temperature, sulphide inclusions, which are normally present, are taken into solution by the surrounding metal. On cooling again the sulphides and remaining sulphur precipitate out and tend to segregate to the grain boundaries as liquid films, thus weakening the grain boundaries considerably. Steel which is embrittled in this way is referred to as burned.

During welding of steel a region of the HAZ adjacent to the fusion boundary is heated to a temperature which would normally cause burning. If the heat input of the welding process is low (low current fast travel) the HAZ is narrow, and burning may be hardly discernible. When the heat input is large, because of high currents in submerged arc welding or slow welding speed in oxyacetylene welding, the burned region of the HAZ will widen. The thermal strains which accompany welding may cause the sulphide films to open up while they are still liquid into small cracks or "hot tears".

The susceptibility to burning and hot tearing appears to depend mainly on the relative amounts of sulphur, carbon and manganese in the steel although other elements such as phosphorus and certain alloying elements may be important. As a general guide, a manganese to sulphur ratio of at least 20:1 is necessary to avoid burning during metal arc welding of steel with a carbon content of about 0·20%. With higher carbon content or with the use of high heat input processes the Mn:S ratio may need to be above 30 to give freedom from hot tearing and as high as 50 to reduce the extent of burning.

The practical implications of burning and hot tearing will be discussed further under the heading 'Factors affecting joint efficiency'.

LAMELLAR TEARING

Another problem which is largely associated with segregated impurities in the parent material is that of lamellar tearing. This problem occurs particularly when making tee and corner joints such that the fusion boundary of the welds runs parallel to the plate surface and tensile residual stresses can act across the plate thickness. Cracking often occurs in the HAZ but may occur well away from it. The crack path is generally step-like in character with comparatively long straight portions parallel to the plane of the plate.

The basic cause of cracking appears to be inclusions such as sulphides and silicates which are rolled out into planar form during manufacture of the plate. These inclusions result in lower ductility and sometimes reduced static strength in the short transverse or through thickness direction of the plate. Under the action of welding stresses, particularly where a high degree of restraint is involved, the planar inclusions may open up and run together to form a crack. It is not necessary for the impurities to be present in the form of large sheets but often a large number of small areas in slightly different planes. This gives the crack its characteristic step-like appearance. It should be pointed out that, unlike laminated plate, these planes of weakness cannot be detected, with current techniques at least, by ultra sonic or other forms of inspection.

The susceptibility of steel to this type of defect increases with increasing plate thickness. This results from the possible increase in composition limits and higher degree of segregation associated with thicker sections and from the higher level of restraining forces involved.

Lamellar tear adjacent to T butt weld in cruciform joint

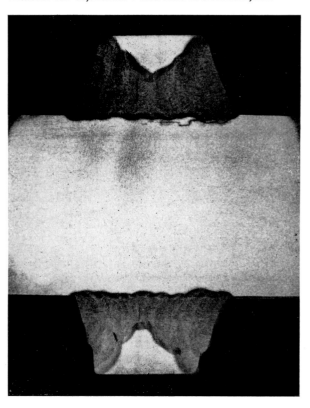

Lamellar tear in a mild steel nozzle at the nozzle to shell circumferential T butt joint

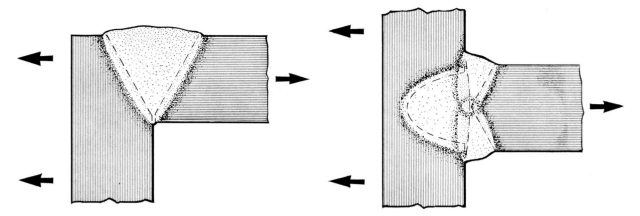

To reduce the risk of lamellar tearing consideration should be given to the design of joint details particularly where high restraint during welding is unavoidable. Welding procedures should be such as to reduce the effects of restraint where possible. Edge preparations for corner joints should be designed so that the fusion boundary of the resulting weld runs across the possible planes of weakness of the material. For tee joints the base plate may be gouged out and then filled with weld metal before making the actual joint weld. This technique is expensive and should only be used where practical experience with the particular material and joint configuration involved shows it to be necessary.

HEAT AFFECTED ZONE HYDROGEN INDUCED CRACKING

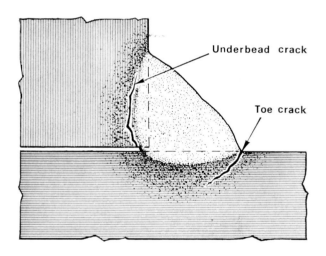

Heat affected zone cracking, hydrogen cracking, hard zone cracking, cold cracking, toe cracking and underbead cracking are all different terms used to describe the same phenomenon. The fact that so many terms are in fairly common usage indicates the widely occurring nature of the problem. Each of these terms describes a particular facet of the problem.

These cracks occur in a hardened HAZ adjacent to and frequently parallel with the fusion boundary. The susceptibility to cracking is increased by the presence of hydrogen and the cracks occur spontaneously after the joint has cooled to a temperature of about 300°C or lower. It may be several hours after welding when cracking takes place.

Typical underbead crack adjacent to a fillet weld

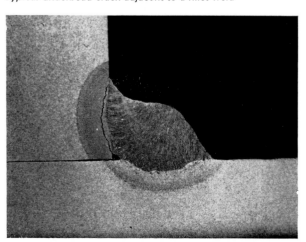

Toe cracks in the heat affected zone of a butt weld

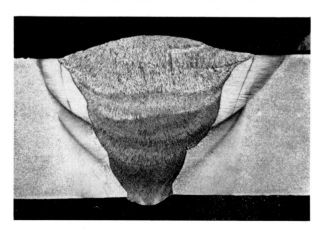

Four of the main factors which affect the incidence of heat affected zone cold cracking are:—

the thickness of material and type of joint,

the composition of the steel,

the type of electrode or welding process used,

the energy input of the welding process and the preheating temperature.

These variables are outlined and it will be seen that they are to a large extent dependent upon each other.

THICKNESS AND TYPE OF JOINT

It is well known that many steels may be hardened by heating to a particular temperature followed by rapid cooling and that the resulting hardness may be varied by changing the rate of cooling (*e.g.* by quenching in water or oil). It is this phenomenon which operates in the HAZ of a weld and produces a hardened microstructure. For a given steel composition the hardness of this microstructure will depend on the cooling rate in the HAZ. Variations in the thickness of material and in the type of joint result in modification of the cooling rate in the HAZ and of the residual stress system associated with the weld. Most of the heat abstracted from the HAZ is conducted into the plate material. In general, for a given heat input, the thicker the plate, the more rapid the rate of cooling. The overall size of the components is also important but in most cases in practice, the members being joined are relatively large and can be considered to provide an infinite heat sink. In other words if the amount of heat put into the joint during welding was spread uniformly throughout the components then the temperature of the material would not be raised significantly.

The type of joint is significant in that it governs the number of paths along which heat may be conducted away from the weld. The rate of cooling in the HAZ of the root run in a V butt weld, for example, will be much less than that in the root run of a fillet welded tee joint for the same plate thickness, heat input and initial plate temperature. It should also be noted that the edge preparation may affect the initial cooling rate to a considerable degree.

As indicated above the residual stress system associated with welding also has an important effect on the likelihood of cracking. The level of stress in the HAZ will depend to a large extent on the degree of restraint afforded by the surrounding material. Restraint, like ductility and weldability, is a relative term with no absolute definition, but in this context it may be defined as the resistance of the joint configuration to deformation and distortion of a kind that would relieve the welding stresses. The amount of restraint will, in general, increase as the thickness of members is increased.

Comparison showing the effect of root gap on the susceptibility to heat affected zone cracking.
Top: $\frac{1}{64}$ in. gap; Bottom: $\frac{1}{16}$ in. gap

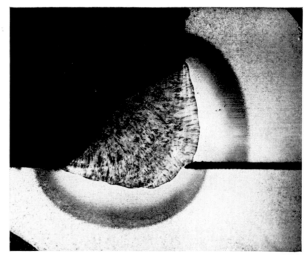

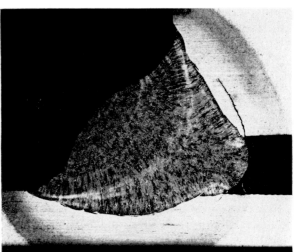

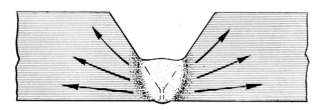

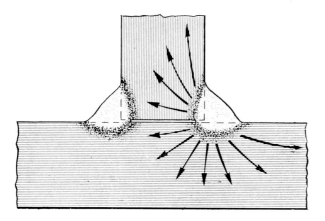

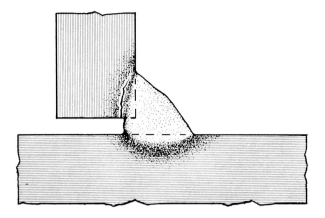

There is a further factor which should be considered. Evidence exists both from laboratory tests and practical examples that, even with low restraint, cracking may occur when conditions encourage the local intensification of strain. The effect of a root gap, caused by bad fit up, in fillet welded components, for example, is to markedly increase the possibility of heat affected zone cracking, even though all other procedural variables are kept constant. The reason for this may be that concentration of stress occurs at the geometrical notch existing between the weld metal and parent plate at the root of the weld as it cools. A similar effect may occur in any incompletely penetrated weld. It has also been found that the likelihood of cracking adjacent to the last run in a partially completed butt weld which has been allowed to cool, is greater than that adjacent to the last run of an identical completed weld.

These examples show the necessity for giving attention to the choice of welding procedure to reduce the effects of restraint, bad fit up and unsuitable edge preparations.

COMPOSITION OF THE STEEL

With a given welding procedure, joint configuration and plate thickness, the likelihood of cracking will depend entirely on the composition of the steel since this determines the type of microstructure in the HAZ. It is not proposed here to go into metallurgical detail on the different types of microstructure, or transformation products, which may be produced by various steel compositions under the action of rapid cooling conditions. There are two main aspects of the transformation behaviour of steels which are important :—

The hardenability of the steel,

The susceptibility of the hardened structure to cracking.

Hardenability is the ease with which hardened heat affected zones may be formed. In other words a material with low hardenability will require very rapid cooling rates to produce a hardened microstructure, whereas a material with high hardenability will transform to a hard HAZ even at very slow cooling rates. Even though a hard HAZ is formed it still may not be susceptible to cracking. Both hardenability and susceptibility are increased by raising the carbon and alloy content, although certain alloying elements increase the hardenability without a significant increase in susceptibility as compared with carbon. The dominant effect of carbon has been recognised by the production of various carbon equivalent formulae which attempt to assess the effect of composition, but it should be pointed out that they may only apply to the particular compositions used in determining them and are certainly not universally applicable.

One of the reasons for the limited success of carbon equivalents is that commercial steels are not characterised by a single chemical composition. As mentioned previously it is necessary in producing steel to a particular specification to have fairly wide limits on composition to produce the same mechanical properties over a wide range of thickness. Segregation of some of the elements during solidification may give an additional variation in composition from one part of a plate to another. Both limits of composition and possible segregation effects tend to increase with greater plate thickness. Since HAZ cooling rates also increase with thicker sections it can be seen that the possibility of producing susceptible microstructures is much greater. It is possible, for example, to get cold cracking in mild steel in thicknesses of 2 inches, although greater thicknesses are usually required.

TYPE OF ELECTRODE OR WELDING PROCESS

The type of welding electrode or process is one of the most important variables affecting the likelihood of HAZ cracking. The susceptibility of a hardened microstructure to cracking is influenced to a large extent by the amount of hydrogen introduced into the weld by the welding process. The precise mechanism by which hydrogen increases the susceptibility is not fully understood but it has been shown that the fracture strength of hardened steel is drastically reduced by the presence of as little as 2 ml. of hydrogen per 100 grms. when tested at ambient temperature. As the temperature is raised the embrittling effect of hydrogen diminishes.

If the steel being fabricated does not harden under the rates of cooling produced by the process being used, then the control of hydrogen content is not so important. Class 2 or 3 titania covered (rutile) electrodes producing weld metal hydrogen contents of 20 to 30 ml. per 100 grms. may be perfectly satisfactory for mild steel structures up to $1\frac{1}{2}$ inches thick. With the increased use of higher tensile low alloy steels, the control of hydrogen content becomes much more important. The use of hydrogen controlled electrodes giving weld metal containing less that 10 ml. of hydrogen per 100 grms. is usually essential and as the material thickness is increased so additional precautions may become necessary. The argon and CO_2 gas shielded metal arc processes can produce weld metal deposits having considerably lower hydrogen concentrations than the hydrogen controlled covered electrodes. This does not mean that no precautions are necessary with these processes. It must be emphasised again that significant amounts of hydrogen may be introduced into the weld if either joint or electrode wire are contaminated with oil, dirt, rust or moisture.

ENERGY INPUT AND PREHEATING TEMPERATURE

Another important variable controlled by the welding process is the energy input to the joint. For a given plate thickness, the greater the energy input per unit length of weld, the slower the cooling rate will be in the HAZ. The heat input is a function of current, arc voltage and welding speed. The main variables are current and speed of travel. To increase the heat input it is necessary to increase the current or reduce the travel speed. For manual metal arc welding this generally means using larger electrodes and in steels of relatively low hardenability it may be possible to produce an unhardened HAZ simply by increasing the electrode size. Even if the HAZ transforms on cooling the tendency to cracking will be reduced by a slower rate of cooling because of the lower hardness of the resulting microstructure.

There are practical limitations to the amount by which the energy input may be increased and it is frequently necessary to resort to other procedures to reduce cooling rates. Preheating is the most usual technique and may involve heating the material to temperatures up to 350–400°C. The level of preheat required in any particular case will depend on the composition and thickness of material, the type of joint and welding process used. Preheating, like energy input, may not in many cases affect the transformation product but the slower rate of cooling after transformation has an additional benefit in that it may allow hydrogen more time to disperse from the joint and thus reduce the sensitivity to cracking.

On preheated joints requiring several runs of weld metal it is essential that the temperature does not fall below the preheat temperature and in some cases it may be necessary to apply heat additional to the welding energy to maintain this interpass temperature.

Since HAZ cracking depends on so many factors it may be useful to summarise the general precautions to be taken in relation to particular types of steel. It is convenient to classify structural steels into groups according to their hardenability, on the one hand, and the nature of the hardened structure on the other. Five general classifications are shown in the table with summaries of the effective precautions for avoiding cracking. These are further discussed below.

Class 1 steels

These steels do not generally produce hardened structures under normal metal arc welding conditions, and no special precautions are normally required to prevent HAZ cracking. When welding very heavy sections (*e.g.* 4 in. or greater) it may be

Class	Description	Welding precautions	Examples
1	Unhardened under normal metal arc welding conditions.	None normally required.	Mild steel (<·15%C, <·8%Mn) and *e.g.*, Corten and Fortiweld.
2	Low hardenability with low susceptibility when hardened.	Low hydrogen electrodes recommended. Preheat may be necessary under the more severe conditions, *e.g.* thick sections or small electrodes	Mild steel (·15%<C<·25%, <1·0%Mn) C – Mn steels (⩽·2%C, ⩽1·4%Mn)
3	Low hardenability with high susceptibility when hardened.	Preheat may be unnecessary using high heat input (large electrodes). If preheat is necessary it should generally be in the range 250°–350°C.	C – Mn steels (>·25%C, ⩽1·0%Mn), *e.g.*, En 5, 6, 8 and 9.

Class	Description	Welding precautions	Examples
4	High hardenability with low susceptibility when hardened.	Low hydrogen electrodes recommended. Preheat and interpass heating will generally be necessary, the level depending on composition, thickness and technique.	Most low carbon, low alloy high strength steels, *e.g.*, BS. 1501/151 and 161, BS. 968 :1962 (>1 in. thick) Steels up to ·15%C, 1·5%Mn, 1·5%Ni, 1·0%Cr, ·25%Mo, ·2%V.
5	High hardenability with high susceptibility when hardened.	Preheat and interpass temperature in the range 150° – 250°C are necessary, together with immediate post weld softening heat treatment.	Alloy steels with C > ·25% except as defined in classes 1 – 4, *e.g.*, En 15 – En 30 inclusive.

necessary to use low hydrogen electrodes or a preheat to be sure of preventing cracking. Occasionally problems may arise at thicknesses less than this where a high degree of segregation is evident. Welding on the cut edges of plate where the cut passes through particularly heavily segregated areas of the original ingot.

Class 2 steels

The likelihood of cracking in these steels is reduced if the hydrogen content introduced during welding is kept as low as possible. Low hydrogen electrodes or a low hydrogen process are recommended. Increasing the heat input of the process will be beneficial, since it may produce a softer HAZ microstructure and allow more time for hydrogen to diffuse out of the joint during cooling. For welding thick sections or where small electrodes have to be used, preheating may be necessary and this has roughly the same effect as a high heat input in reducing the rate of cooling.

Class 3 steels

For these steels, which have a high susceptibility towards hydrogen cracking, it is unwise to rely solely on reduction in hydrogen concentration and instead, the object should be to achieve a non-hardened HAZ. It may be possible to avoid preheat if the process heat input is sufficiently high. Preheat and interpass temperatures up to 350°C may be necessary in addition to achieve a softened HAZ.

Class 4 steels

The hardenability of such steel is generally too high for preheat alone to produce a soft HAZ. The susceptibility to cracking is not unduly high, and so a hardened structure can be tolerated, providing the hydrogen content can be sufficiently reduced before cooling to normal temperatures. Low hydrogen electrodes or a low hydrogen process must always be recommended and some degree of preheat and interpass heating may be necessary depending on the composition, thickness and type of joint welded. The effect of preheat is not primarily to alter the structure in the HAZ, but to reduce the cooling rate so that more hydrogen can diffuse from the joint before it cools to ambient temperature.

Class 5 steels

The HAZ microstructure of these steels is so highly susceptible to cracking that, except with very thin sections, it is necessary to produce a soft HAZ before the weld is allowed to cool to normal temperature. Preheat must be maintained as an interpass temperature during welding and an immediate post weld heat treatment applied. This may be achieved in one of two ways:

By using a high preheat and then maintaining the welds at that temperature after welding for a sufficient time to allow transformation to a softer structure.

By maintaining a lower preheat and interpass temperature and then raising the temperature immediately after welding, thus tempering the hard HAZ to a softer structure.

It is not possible to set out the precise conditions required for welding a particular steel. It is usually desirable to carry out procedural tests to define these conditions. Procedural tests should be designed to simulate the most unfavourable combination of variables likely to be encountered during fabrication. Tests should be done on joints involving the thickest sections with the heat sink at least equivalent to that in the actual construction and any misfit of components should be the maximum allowed. If preheating is required this should entail heating the minimum area to the minimum temperature to be specified for the actual construction. Tests of this type will not only check that the chosen welding procedure is satisfactory but also establish an adequate margin of safety in respect of normal practical variables. It is not sufficient to look for cracks large enough to be evident in ordinary visual inspection. The parts must be sectioned and examined under a microscope. Also time must be allowed before sectioning for any cracks to develop.

HEAT TREATMENT CRACKING

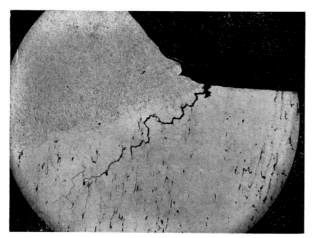

Post weld heat treatment crack in the heat affected zone in austenitic stainless steel initiated from a small hot tear at the weld toe

Heat treatment may be necessary on a welded fabrication for a variety of reasons: for example, the relief of residual stresses to give dimensional stability for subsequent machining operations; or the restoration of weld metal or HAZ properties to give greater protection against brittle fracture or corrosion. Some fabrications may be subjected to a service life at elevated temperature and in this context may be regarded as undergoing a low temperature heat treatment. Certain steels, particularly austenitic stainless steels and ferritic creep resisting steels, are susceptible to embrittlement during heat treatment which may give rise to cracking in the heat affected zones of welded joints.

Good creep strength at high temperatures in steel is usually derived from a finely dispersed precipitate of carbides throughout the metal grains. During welding of such steels, the material in the heat affected zones undergoes a high temperature solution treatment which means that the carbides are dissolved and taken into solution by the metal. When the weld cools some of the carbides are precipitated but due to the high rate of cooling some of the elements are retained in solution. This means that the material is unstable and if it is reheated, further carbide precipitation can occur. The residual stresses present in welded structures are relieved by creep deformation during reheating. This creep relaxation occurs by a combination of grain boundary sliding, *i.e.* one grain moving relative to another, and grain deformation. If, during reheating, the carbides are precipitated as a fine dispersion within the grains then the high temperature strength of the grains is increased and consequently the creep ductility is reduced. This in turn, means that most of the relaxation of residual stress has to occur by grain boundary sliding and if the grains themselves do not deform, can result in the boundaries opening up into cracks.

Before cracking can occur there must be a sufficiently high stress level present irrespective of the state of embrittlement of the material. Thick sections are particularly susceptible because of restraint. Welding procedures designed to reduce restraint will be beneficial. The use of weld metal having a lower hot strength than the parent plate will be useful as this will allow more relaxation to occur in the weld metal than in the HAZ during reheating.

A sound welding procedure is essential to ensure that the joints are as free as possible from defects such as slag inclusions, small cracks, etc. These may have the effect of locally raising the level of stress. It is generally good practice to grind the toes of welds before heat treatment to remove any small surface hot tears which are quite likely to have occurred during welding.

Tests have shown that the degree of embrittlement occurring during heat treatment can be controlled by the temperature at which the treatment is carried out. At low temperatures the precipitate consists of a large number of small particles dispersed throughout the grains whereas at higher temperatures the carbide particles are larger and fewer in number. The coarser precipitate gives a structure which is less hard and has a higher creep ductility. It is thus better able to accommodate the plastic strains necessary to relieve residual stresses. Heat treatment should be carried out at the highest temperature for the particular steel in question. For the ferritic steels this will usually be between 650°C–700°C while for the austenitic steels 1050°C will be necessary. It is important that the rate of heating should be as fast as possible to avoid any appreciable precipitation at the lower temperatures during reheating. A practical limit is imposed by the danger of inducing severe thermal stresses if the heating rate is too rapid.

Apart from the problem of precipitation embrittlement in certain types of steel, service at elevated temperatures may also give rise to creep cracking, particularly in the weld metal. Creep strain is the slow plastic deformation which can occur under the action of stress and increasing temperature. The amount of strain in a given material will depend on the magnitude of these two variables. The creep ductility of weld metal is usually much lower than that of the parent material so that even at relatively low average strains cracks may occur in the weld metal. These cracks usually appear transverse to the line of the weld and may propagate into the HAZ. The presence of residual stresses will significantly increase the risk of creep cracking and lower the temperature at which this can occur. Structures which are required to operate at elevated temperatures should generally have a stress relieving heat treatment after welding.

CORROSION

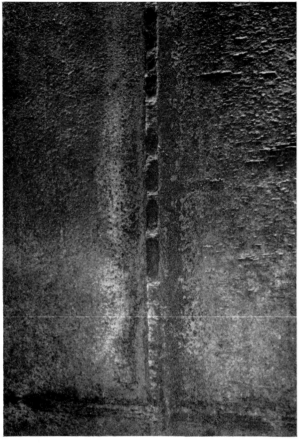

Weld metal attack caused by paint breakdown (with acknowledgments to Mr. S. G. Christennssen, Supt. Engineer, Sugar Line Ltd.)

Corrosion is a large and complex subject and its effects on steel structures, whether or not they are welded, can be drastic. Welding can introduce particular problems which are described in this section.

The mechanism of corrosion is generally electrochemical and the formation of electrolytic cells at metal surfaces can occur in many ways. The basic requirements for this to take place are a liquid in contact with the metal and a difference in electric potential between two points. The liquid may be moisture condensed from the atmosphere or liquid contained in a vessel. The potential difference may arise from two different materials being in contact or local differences in composition of the liquid from one point to another. If suitable conditions exist then two reactions are set up. At certain points or anodes an 'anodic reaction' takes place in which charged particles of metal or 'ions' enter the liquid leaving behind free electrons. This reaction can continue only if elsewhere a 'cathodic reaction' has been established to use up these free electrons. The cathodic reaction will depend on the nature of the liquid but, for example, may be the reduction of oxygen to hydroxyl ions or the liberation of hydrogen gas from the liquid. The flow of current or electrons, in a corrosion cell results in attack on the metal at the anodes but leaves the cathodes uncorroded. The overall rate of corrosion will depend on the metal involved and on the particular environment but the ratio of anodic area to cathodic area is also of great significance in a corrosion cell. The smaller the area of anode relative to the cathode the greater the rate of penetration at the anodic area.

A dissimilar metal joint immersed in a liquid is perhaps the best known example of the formation of an electrolytic cell. This can occur in welded joints if the composition of the weld metal forming the joint is significantly different from that of the parent plate. The variations in composition between weld metal and parent plate in normal steel structures are not usually important unless the environment is particularly severe. If this situation does arise, it is usually the case that the weld metal forms the anode and because of its much smaller area, is rapidly corroded away. The remedy under these conditions is to ensure that the weld metal composition matches the parent material as closely as possible.

Millscale on the surface of plate becomes cathodic to the steel so that any breaks in the scale form anodes and the steel is attacked. At welded joints the millscale is removed by the heat of welding and if the remaining scale is left intact then again the welded joints may be preferentially corroded. Welding slag which is not completely removed from the joint can have a similar effect but in this instance the relative areas would be such that only a smal degree of corrosion could occur. Small particles of slag can react with protective paint films which may be applied and cause these to break down locally. This then exposes small areas of steel which become anodic to the rest of the surface and may give rise to deep pitting at these points. Problems of this kind can be prevented by attention to cleaning operations prior to the application of any protective coatings.

Joints containing crevices which can trap and retain moisture are another source of concentrated corrosive attack. A butt joint made on a permanent backing strip is a typical example.

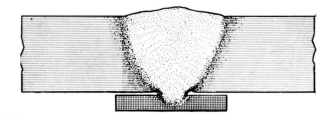

The electrolytic action under these conditions is usually maintained by a difference in composition of the liquid inside and outside the crevices. The root of the crevice usually forms the anode and hence concentrated attack occurs at this point. It is not always possible to eliminate details of this kind at the design stage but every effort should be made to do so, particularly where severe corrosive environments are likely to exist.

Microstructural changes in the HAZ of certain materials may give rise to particular problems. One of these which has become fairly well known is that of 'weld decay' at joints in austenitic stainless steels. In this particular case a region in the HAZ is heated to a temperature of 550°C – 850°C during welding. Within this temperature range, carbon which is in solution can precipitate out as chromium carbide at the grain boundaries. Because of this carbide precipitation a region round each grain is left deficient in chromium and it is this depleted region that is attacked by some corrosive media, resulting in disintegration of the steel into individual grains.

Heat treatment after welding at a temperature of 1050°C followed by rapid cooling will remove the carbides back into solution and restore the corrosion resistant properties. This 'solution heat treatment' is not always practicable, nor particularly economic. For this reason 'stabilised' steels were developed which contain either niobium or titanium. These are both strong carbide forming elements and form carbides in preference to chromium over most of the range of temperature in which precipitation can occur. The chromium is retained in solution, maintaining the corrosion resistance of the material. A region of the HAZ of stabilised steels may become susceptible to intercrystalline corrosion, through chromium carbide precipitation, if the material is heat treated following welding. Where heat treatment of stabilised steels is required, as a stress relieving treatment for example, a temperature of 850°C or higher is recommended to avoid precipitation of chromium carbide.

As a result of the possible drawbacks associated with stabilised stainless steels the use of extra low carbon varieties is becoming more attractive in spite of their greater cost. If the carbon content of the steel is below the minimum necessary for precipitation during welding, usually less than 0·03%, then the corrosion resistance is not impaired by welding or subsequent heat treatment.

Another effect of corrosive environments, which applies particularly to welded structures, is that of stress corrosion cracking. This is a phenomenon which is encountered in many materials, both ferrous and non-ferrous. As its name suggests, this phenomenon requires the material to be in a state of tensile stress and in contact with a corrosive medium. The level of stress necessary to cause cracking may be well

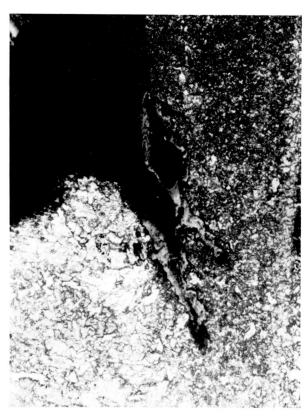

Intergranular corrosion or 'weld decay' in austenitic stainless steel

Corrosion at the root of a crevice in a welded mild steel pipe joint

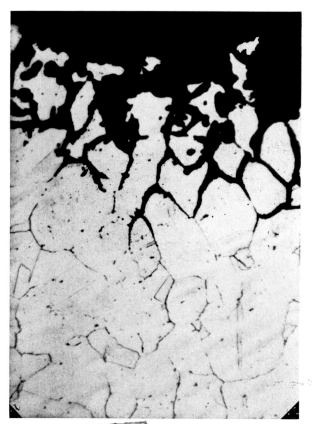

below the yield point of the material depending on the environment. Some materials are only sensitive to a particular impurity, whereas others may be sensitive in some degree to several impurities. Many stainless steels, for example, are particularly susceptible to chlorides or hot caustic solutions. Mild steels may be susceptible in environments containing nitrates, particularly at above ambient temperatures, and to boiling caustic solutions.

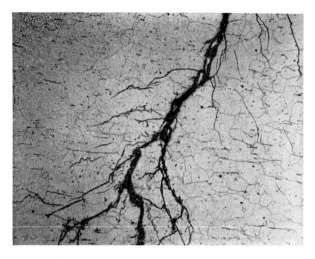

Transgranular stress corrosion cracks in austenitic stainless steel

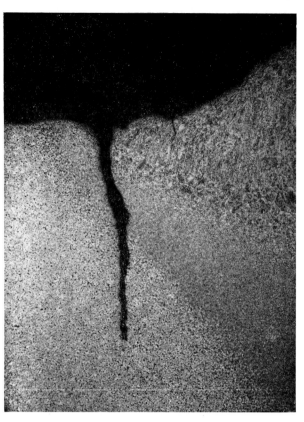

Typical corrosion fatigue crack in mild steel initiated at the toe of a fillet weld

Stress corrosion is a difficult problem to predict as, in some instances, the concentration of impurities sufficient to cause cracking may be as low as a few parts per million. Structures which are likely to give trouble should be stress relieved where possible to reduce the welding stresses. It is sometimes necessary to design to lower working stresses to minimise the risk of cracking. Any crevices, notches or sharp changes in section should be avoided as these will not only give rise to local intensification of stress but may also allow concentration of the impurities likely to cause cracking. Unfortunately, the problem usually becomes apparent only during service when any remedial measures to reduce the stress level or change the environment can be very expensive.

The last problem to be discussed under the general heading of corrosion is that of corrosion fatigue. Structures which operate under cyclic or fluctuating loads may eventually suffer a fatigue failure. The likelihood of fatigue cracking will depend on the magnitudes of the fluctuating stresses, the severity of any stress concentration such as notches, discontinuities or changes of section particularly at welded joints, and the number of fluctuations of stress occurring during service. For a given notch or welded joint the magnitude of stresses necessary to cause cracking reduces as the number of repetitions of stress increases. In other words, the longer the required life of the structure then the lower must be the working stresses. When fatigue loading and a corrosive environment are combined, the resulting problem becomes more severe than when the two effects are separate.

If the fluctuating stresses in a structure are high enough to initiate small cracks at stress concentrations then anodic reactions may be set up within these cracks and cause them to propagate more rapidly. Alternatively the stresses may be so low that fatigue cracks would not normally occur, but the corrosion may be more severe and at regions of local attack on plate surfaces cracks may initiate and propagate, even at the lower stresses. Once cracks have formed this will have the effect of concentrating the corrosive attack still further. In general terms the problem extends between the two extremes of fatigue accelerated by corrosion and corrosion accelerated by fatigue. The extent to which the problem can be controlled will depend on which of the two variables is of prime importance in any particular case. Unlike stress corrosion which requires a specific environment, corrosion fatigue may occur where fatigue loading and any corrosive condition exist. As with other aspects of corrosion and fatigue, attention to detail design in eliminating notches and discontinuities of section is a first essential.

FACTORS AFFECTING JOINT EFFICIENCY

Various problems and defects which can arise as a result of welding have been described. How these defects will affect the performance of a joint will depend on the particular conditions of loading and environment to be imposed.

The efficiency of a joint may be defined in many ways: for example, in terms of static strength, fatigue strength, fracture toughness or ductility at low or elevated temperature. Since the designer is usually concerned with the strength performance of joints, the effect of welding deficiencies on the structural properties are further described.

STATIC STRENGTH

In mild steel and low alloy steels, where strength is derived from the presence of carbon, the static strength of a welded joint is normally greater than that of the parent material. This is because both the weld metal and HAZ are rapidly cooled from a high temperature and transform to microstructures of comparatively high strength. They are thus stronger than the parent metal, which is generally in the as-rolled, normalised, or quenched and tempered condition. For this reason mild steel weld metal can be used for joining steels of up to 30 tons/in^2 ultimate tensile strength without special alloy additions.

Under purely static loading conditions and at temperatures where there is no risk of a brittle failure (this will be described later), defects can only lower the strength in proportion to the area of cross section they occupy. Where the strength of the weld metal is higher than the parent metal, as explained above, even extensive areas of defectiveness will not reduce the strength of the joint below that of the parent plate. This also applies even at high rates of loading, such as impact to which many engineering structures are subjected, unless the loading is sufficiently repeated to come into the context of fatigue.

Static loading can produce failure in welded joints when combined with other factors and in these circumstances defects become more important. One can consider, as an extreme example, forming or flanging operations subsequent to welding. In this instance a high degree of ductility may be required in the welded joints. If the ductility of the weld metal or HAZ is significantly lower than the parent material, cracking may result. Any defects, such as hot cracks in the weld metal or cold cracks in the HAZ will extend under the high strains imposed by forming operations.

Creep rupture and stress corrosion are also examples of failure under static loading in combination with elevated temperature and corrosive environments respectively. These have been touched on briefly under the headings of 'heat treatment cracking' and 'corrosion' and will not be further described except to say that any welding defects which are present may serve as initiation points for more severe cracking during service.

BRITTLE FRACTURE

A brittle fracture or cleavage fracture is a sudden failure accompanied by little or no visible yielding of the material. It is in relation to the risk of brittle failure that defects and poor joint properties become really important.

Before going further it may be useful to outline the conditions under which brittle fractures can occur. Three factors are necessary:

A temperature below the transition temperature of the material.

A notch or severe stress concentration.

A tensile stress.

The transition temperature of a steel is the temperature or range of temperature over which the mode of fracture of the material when notched rapidly changes from ductile to brittle. Different steels have different transition temperatures, but the same steel in different conditions and thicknesses may also show wide variations in transition temperature. A steel which is normalised, for example, will generally be tougher, *i.e.* have a lower transition temperature than the same steel in the same thickness in the as-rolled condition. This is caused by the much finer grain size of the material resulting from the normalising process. The transition temperature for different steels is usually determined by the Charpy V notch impact test in which the energy for fracture is measured for a series of small notched specimens tested over a range of temperatures. Correlation should be established between these small scale test results and the full thickness service behaviour of the same steel to establish a Charpy energy absorption level at which a high expectation of freedom from brittle fracture may be implied at the required minimum temperature. This correlation is extremely difficult and is the subject of considerable research.

Materials below their transition temperature are not necessarily weakened. This is why a notch or stress concentration is a requirement for the initiation of a brittle fracture. This also explains why, strictly speaking, one talks in terms of 'notch toughness' or 'notch brittleness' as properties of materials. The notch has several functions: it produces stress concentration and, particularly at yield stress and above, a concentration of strain; it inhibits ductile slip of the material grains by producing biaxial or triaxial tension stresses in the plane of the notch. This reduction in ductility and accumulation of stress, favour brittle facture.

No designer intentionally introduces notches into a structure, but unfortunately welds may produce far more severe notches than many designers realise. This is perhaps best illustrated by considering the fatigue strength of a machined member which is reduced to 1/5th of its previous value when a fillet weld is made on it. In a particular set of circumstances the severity of notch necessary to initiate a fracture will depend on the sensitivity of the material in which the notch is situated.

To produce any fracture, a tensile stress is required in the material. In welded structures, this tensile stress may be the residual stress due to welding, an applied tensile stress, or more usually a combination of both. The material does not distinguish between applied and residual stresses, and if the magnitude of the stress is high enough and the other necessary conditions for fracture are present, then a brittle failure will occur.

It is more difficult to initiate than to propagate a brittle crack. A stress of a few tons/in^2 is sufficient to enable the crack to propagate but, in general, yielding at the notch is required to initiate fracture. This does not mean that yield magnitude stresses are required over a large area, as the stress may be intensified locally in many ways. A low stress fracture implies that the structure has failed at an applied stress lower than the design stress but, nevertheless, yield point or higher stresses will have occurred at the tip of the notch.

The effect that weld defects and deficiencies may have on the problem of brittle fracture can now be described in more detail. It should be clear that hot or cold cracks, hot tears and other crack-like defects such as lack of penetration and lack of fusion will form severe stress concentrations and are all potential fracture initiation points. The majority of welding defects are situated in a high tensile residual stress field so that at the transition temperature of the material surrounding the defect, little or no additional stress may produce a fracture. What is perhaps not so obvious is that during or after the formation of a crack-like defect, the material at the tip of the crack will be plastically strained at an elevated temperature because of the welding process. This material may be strain aged, or with some steels precipitation embrittled, reducing the notch toughness compared with the parent plate. Once a fracture has initiated from a defect such as this it may well continue to propagate even where the fracture toughness of the parent material is high.

The notch toughness of the HAZ of welded joints may be severely reduced by microstructural changes which are brought about by the weld thermal cycle. One of the major reasons for a fall in toughness in structural steels is the possibility of burning in the HAZ. The causes of burning and hot tearing have already been explained but if hot tears or any additional defects exist in the burned region, then the notch toughness will be much reduced. Where the other conditions of stress and low temperature occur during service there will be a brittle fracture risk.

Fortunately the risk of brittle fracture may be eliminated by removing any of the three conditions necessary to cause it. That is to say, at a service temperature above the transition temperature, even fairly large crack-like defects can be tolerated and the problem becomes one of static strength only. Alternatively, even if the material is brittle at the service temperature there is no risk of fracture provided that there are no defects present. This is a difficult condition to achieve, even with very good welding procedure, and for this reason is not very practical.

Defects may be tolerated in material which is brittle if the structure has undergone a stress relieving heat treatment subsequent to welding in order to eliminate the residual welding stresses. Brittle fracture under these conditions will normally require applied stresses in the order of general yield of the material and these will not occur in a structure unless a serious mistake in design calculations has been made. Stress relieving has an additional benefit in many cases by removing local metallurgical damage such as strain ageing, for example, at the tip of existing weld defects. Stress relieving will not improve the notch toughness of the parent material and so will not increase its resistance to fracture propagation. Therefore it is important that no welding be done on a structure after it has been stress relieved. Some materials, particularly certain low alloy steels, may suffer a reduction in notch toughness because of precipitation effects during heat treatment and so may become more susceptible to fracture than in the as-welded state.

On large fabrications, thermal stress relieving may be impractical. Local stress relief by means of blankets, etc., needs to be approached with care, as, if the heated area is not large enough, new residual stresses will be induced on cooling. Under these conditions, it is important to select the material which gives the necessary level of notch toughness at the appropriate service temperature required in the as-welded state.

FATIGUE STRENGTH

Fatigue failure is probably the most common type of failure in welded construction, but it is still not widely appreciated that the fatigue behaviour of a welded structure depends on the fatigue behaviour of the welded details composing it, not on the fatigue properties of the parent material. The majority of fatigue failures experienced in practice stem from bad design and a lack of knowledge of the severe stress concentrations which may be introduced in welded design.

To take a simple example, the fatigue strength of a welded mild steel plate girder with a continuous web to flange weld for a life of two million cycles will depend on whether the weld is made by continuous automatic welding or by manual welding with its inevitable stop-start points when the electrode is changed. As a result of the stop-start points the fatigue strength of the manually welded girder will be about 9 tons/in^2 in terms of the maximum bending stress in the flange and that of the automatically welded girder with no stop-start points will be 12 tons/in^2. A transverse butt weld in the flange may reduce the fatigue strength to 7 tons/in^2 while transverse fillet welds attaching a stiffener to the flange will reduce the strength still further to about 5 ton/in^2. These figures assume that there are no serious defects in the joints such as cracks, lack of penetration, large slag inclusions, etc.

Defects such as these, in so far as they produce stress concentrations, may lower the fatigue strength of welded joints. This does not mean that defects in welds subjected to fatigue loading cannot be tolerated. This example of a welded plate girder should indicate that even quite serious defects could exist in the web to flange weld or the transverse butt weld before a reduction in fatigue strength occurs which is comparable to that produced by the transverse fillet welds.

In assessing the fatigue reducing effect of defects in welded joints it is necessary to consider first and foremost the overall fatigue strength of the structure or component. Fatigue failure will start from the most sensitive point and this may be a small stress concentration in a high stress field or a large stress concentration in a low stress field. The fatigue strength of plain mild steel plate in the as-rolled condition may be about 16 tons/in^2 for two million cycles but the introduction of a transverse butt weld with the normally attendant undercut at the toes of the weld will reduce this figure to 7 tons/in^2. If the weld overfill and undercut are machined flush with the plate surface then the fatigue strength will be restored to 16 tons/in^2 provided that there are no additional defects present. If the joint is used in the as-welded condition and the stress limited to 7 tons/in^2 experiments have shown that quite extensive defects may be present in the joint before the strength is further reduced. Serious defects such as extensive hydrogen cracking in the HAZ of a joint or hot cracks in the weld metal should nevertheless be avoided as far as possible where fatigue loading is concerned.

It can be said that the fatigue properties of welded joints in mild steel are a limitation on the weldability of the material in that, in general, the strength is reduced well below that of the parent plate. This becomes an even more serious limitation of the weldability of the high strength low alloy steels since, except for lives of less than 10^5 cycles, the fatigue strengths of high strength steels when welded are no better than mild steel. The use of low alloy steels often entails a more complicated and critical welding procedure in order to combat other weldability limitations and thus may introduce a higher risk of serious weld defects. For this reason there may even be a higher risk of fatigue failure or possibly brittle fracture, where the appropriate conditions exist.

CONCLUSIONS

This handbook has set out to explain in simple terms many of the problems which may occur during the welding of steels and how these problems may affect the performance of the structure. The majority of these basic problems are very complex and occupy a considerable amount of research work.

It is perhaps inevitable that any description of the problems of weldability may leave a pessimistic impression that welding will involve a high probability of major defects. It should be pointed out that many tons of a variety of different steels are welded daily without giving rise to any of these problems. This is not surprising when it is remembered that most of the difficulties which can arise require the concurrence of several factors. Many fabricators will say that they never have any trouble and even ignore many of the precautions which are normally recommended. What is required is a balanced outlook between this complacent attitude and the over pessimistic view. Provided that the designer is aware of the possible problems which may occur, and the precautions which may be necessary, then he is in a better position to select the correct material. A knowledge of the factors which influence the problem will also enable him to minimise the effect of some of them by suitable design.

It should be stressed that, where difficulties are likely to be encountered in a fabrication, welding procedural trials designed to reproduce as closely as possible all the conditions of the actual detail, will be of great assistance in deciding on the best procedure.

Finally, the Institute's expert staff and facilities are always available to Members to give assistance in cases of doubt and uncertainty.

Printed by LSG Printers, 117-121 Portland Street, Lincoln   Telephone 31631